Thomas Lekscha

# Concept for the Development of a Course of Lectures at Higher Education Level for Training Students of Medical Engineering

**Theme of Course: "Safety in Medical Engineering"**

GRIN Publishing

**Imprint:**

Copyright © 2010 GRIN Verlag, Open Publishing GmbH
Print and binding: Books on Demand GmbH, Norderstedt Germany
ISBN: 978-3-640-74523-4

**This book at GRIN:**

http://www.grin.com/en/e-book/159741/concept-for-the-development-of-a-course-of-lectures-at-higher-education

**GRIN - Your knowledge has value**

Since its foundation in 1998, GRIN has specialized in publishing academic texts by students, college teachers and other academics as e-book and printed book. The website www.grin.com is an ideal platform for presenting term papers, final papers, scientific essays, dissertations and specialist books.

**Visit us on the internet:**

http://www.grin.com/

http://www.facebook.com/grincom

http://www.twitter.com/grin_com

Concept for the Development of a Course of Lectures
at Higher Education Level for Training Students of
Medical Engineering

Theme of Course: "Safety in Medical Engineering"

Thomas Lekscha

# CONTENTS

*Introduction*

This concept is intended to help develop a course of lectures specially aimed at training medical engineering students within the scope of Engineering or Bachelor studies at an institute of higher education.

It should contain and illustrate basic aspects regarding the content of a course of lectures with its emphasis on "Safety in Medical Engineering". This instructional concept should also provide information and procedural instructions on drafting a lecture or lecture manuscript.

*Safety in Medical Engineering*

The term "safety" in medical engineering encompasses numerous, closely linked specialist fields. Safety in medical engineering includes electrical safety, mechanical safety and also chemical and hygienic safety, as well as the safety of medical devices and equipment themselves.

This concept is particularly concerned with the electrical safety of medical devices and medical equipment. The aspect of mechanical safety can be addressed and elaborated on in other lecture courses such as "Materials" or "The use of plastics in medicine".

The subject of chemical and hygenic safety can be addressed and elaborated on in lecture courses related to "Microbiology" and "Chemistry". The aspect of safety within medical engineering is a component part of studies for students of medical engineering and, therefore, should be provided in a series of lectures which have been specifically organized, in didactic terms, for the topic. Parallel to the lectures, laboratory schedules should be drawn up during which the theory conveyed in lectures can be underlined by practical, in-depth experience using real medical devices.

*Aims of the Course*

The lecture course concerning "Safety in Medical Engineering" should have the following objectives and convey them. The course should explain to students of medical engineering how to handle and use medical devices and equipment in a responsible way. When doing so, it is important to include subjects involving the special requirements defined for the development and construction of medical apparatus. Medical engineering students must be made aware of the particular risks which specific medical devices represent and how to prevent these risks by being provided with practical examples. Medical engineering students must be taught how to apply basic legal requirements, specific directives and standards and have the various explanations linked to case studies and real incidents.

After having successfully completed the "Safety in Medical Engineering" lecture course, the students should be capable of assessing practice-related examples with and on medical apparatus. Students should also be able to develop and implement approaches to solutions to prevent safety-related sources of errors.

*Main Legal Regulations*

Since the safety of medical devices encompasses many bordering fields of expertise, only the basic legal regulations, directives and standards should (and can) be addressed during the course of lectures. The most important legal regulations applicable to everyday contact with medical devices are the following:

- Medical Devices Act (German MPG) [1]
  Current version dated 24th July 2010 (BGBl. I S. 983).
  The purpose of this act is to regulate the trafficking of medical devices and, as a result, to ensure the safety, suitability and performance of medical device as well as to ensure the health of patients, users and third-parties is maintained and provide them with the necessary protection.

- Medical Devices Operator Ordinance (German MPBetreibV) [2]
  Current version dated 29th July 2009 (BGBl. I S. 2326).
  This directive applies to the installation, operation, usage and maintenance of medical products in accordance with § 3 of the Medical Devices Act with the exception of medical products for clinical tests or performance evaluation tests.

- DIN EN 60601-1-2; VDE 0750-1-2:2010-05 [3]
  Medical electrical equipment – Part 1-2: General requirements for basic safety and essential performance - Collateral standard: Electromagnetic compatibility - Requirements and tests ( EC 60601-1-2:2007, modified); German version EN 60601-1-2:2010.

- DIN EN 62353; VDE 0751-1:2008-08 [4]
  Medical electrical equipment - Recurrent test and test after repair of medical electrical equipment (IEC 62353:2007); German version EN 62353:2005 Edition: 2008-08

- DIN VDE 0701-0702; VDE 0701-0702:2008-06 [5]
  Inspection after repair, modification of electrical appliances – Periodic inspection of electrical appliances
  General requirements for electrical safety
  Edition: 2008-06, Standard

- DIN VDE 0100-710;2004-06 [6]

  Erection of low-voltage installations - Requirements for special installations or locations - Part 710: Medical locations - Information sheet for the use of (IEC 60364-7-710:2002, modified) Edition: 2004-06.

- BGV A3 [7]

  Accident prevention regulations
  Electrical installations and equipment
  Edition: 2005-01
  Procedural instructions
  Electrical installations and equipment
  Edition: 2005-01.

This list of the most important basic legal texts, documents and standards does not claim to be complete.

However, it does represent a minimal requirement within the development of a concept for a course of lectures concerning "Safety in Medical Engineering".

*Tasks of a Medical Engineer*

To begin the course, the students should be introduced to a detailed description of the occupational tasks for which a medical engineer is responsible. Three professional fields must be presented, accompanied by practical examples:

- Medical engineers involved in development
- Medical engineers involved in sales
- Medical engineers involved in service.

Medical engineers involved in the development of medical devices and equipment must place the emphasis of their studies on the construction and design aspects. Proposes lecture courses will be offered and, in addition, important compulsory and optional subjects will be available.

Medical engineers involved in the sale of medical devices and equipment must place the emphasis of their studies on the business and commercial aspects. In this case, students must participate in inter-departmental lectures with an economic and business content.

Medical engineers involved in customer and technical service in respect of medical devices and equipment must place a great emphasis on the safety-related aspects when considering their professional orientation.

Service technicians must be well-familiar with both the electrical and mechanical safety of medical equipment.

A major objective of a lecture course concerning "Safety in Medical Engineering" should also be to encourage students to disclose their particular preferences and skills (development, sales, service) and to provide assistance and support in selecting the right occupational field.

*Mechanical Safety*

The aspect of mechanical safety should represent about a quarter of the envisaged course time as compared to the aspect of electrical safety. In the case of mechanical safety of medical devices, materials science and the basics of construction technology play a major role. Therefore, a basic condition should be that the students have visited these lectures before participating in the course on "Safety in Medical Engineering". When dealing with the topic of mechanical safety, practical examples on devices should be included in the lecture course.

Key topics which should be expanded on within a course of lectures include: aging of plastics, fatigue fractures, selecting the correct / incorrect production materials, disinfecting materials, ergonomic design of devices and operating errors.

*Electrical Safety*

The electrical safety of medical devices and medical installations should represent about a quarter of the envisaged course time for "Safety in Medical Engineering". A basic condition for participating in this part of the lecture course is students' familiarity with electrotechnical and physical principles and their having visited the relevant lectures beforehand. Since 90% of medical devices and equipment are powered by electricity, it is absolutely essential to expand on the principles of electrotechnology within the "Safety in Medical Engineering" lecture course.

*Current, Voltage, Resistance*

Elaboration of these terms during the lecture course is a basic requirement for understanding the sources of risks of electricity in the actual medical devices and during their use, i.e. both to operators and patients. The lectures should provide a brief but detailed explanation of the origin and creation of an electric current. The difference between direct current and alternating current should be clarified and illustrated by practical examples.

The term "voltage" should be handled in the same way. Important terms which need to be expanded on include: AC voltage, DC voltage, direct current, alternating current, frequency, phases, actual values, crest factor. The term "resistance" has a particularly important significance in the field of safety of medical equipment. The lectures should emphasise the difference between pure ohmic resistance and general impedances. Since the human body does not behave like a pure ohmic resistor, it must be considered according to its specific resistance values when dealing with electrical safety in respect of the use of medical equipment; and parallels or differences to pure electrotechnology or ohmic, capacitive and inductive resistances must be explained.

*Resistances in the Human Body*

In order to expand on the resistance values within the human body, an insight must be provided with regard to establishing the values as well as drawing up and producing resistance tables. Help here takes the form of the initial attempts to establish resistance values from animals (p gs, dogs, sheeps). The transfer and conversion of these experimentally established values to the human body and its peculiarities represents the educational objective. Key topics for elaboration in respect of these resistance values and emphasising the special features of the skin or organ resistance include the structure of human tissue, skin characteristics, current frequency, transition resistance, exposure time, age and constitution of the person.

In order to elucidate the flow of current through the human body or specific bodily impedances, it is helpful to complete a harmless experiment on oneself using a 9 V battery and / or a bell transformer. The students can then discover the various type-related transition resistances on the human body using simple means. In addition, this resource (voltage source, current flow and body resistance) can also be used to recapitulate the Ohm's Law and illustrate it according to a practical example.

*Effects of Electricity on the Human Body*

The lecture course should emphasise to students the dangers and risks which can emanate from electricity (from defective medical devices). Particular attention should be paid to distinguishing between the physical effects and physiological effects of an electric current.

Physical effects:
- Heating due to current flowing through the body
- Boiling of tissue fluids due to high current
- Destruction of proteins
- Burn marks on the body caused by electricity.

Physiological effects:
- Effects on the nervous system
- Muscular cramps
- Interference with the conductive system.

Examples and pictures from the German employers liability insurance association (accidents involving electricity) can emphasise the difference between physical and physiological effects.

*Electrical Safety Precautions*

Electrical safety precautions in the field of electrical engineering serve for restricting unwanted overcurrents (currents which exceed a certain, prescribed threshold value). During the "Safety in Medical Engineering" lectures, students should be presented with the most important electrical safety precautions and safety modules and their relevance highlighted. The basis of these safety features, namely current limitation, should be worked out using example calculations (Ohm's Law and different transitional resistances of the skin).
The following safety features and safety modules should be presented and calculated:

- Safety fuses
  Method of working and different types
  (not reusable after having blown)
- Circuit breakers
  Method of working and different types
  (reusable after having tripped)
- Fault current circuit breaker
  Method of working and different types
  (reusable after having tripped)
- Insulation monitoring module
  Method of working and different types
  (signal only, no shutdown).

Experience has shown that wall charts containing the various type of fuse are very helpful in clarifying safety precautions. Wall charts can show the safety modules as passive components or, when equipped with auxiliary circuits, enable the components to be tripped. A practical illustration of a summation converter or an insulation monitoring module provides support for the theoretical introduction.

*Power Supply Systems (Network Types)*

In order to understand the special features of an electrical power supply in areas used for medical purposes  it is necessary to address the various types of power supply systems and to identify their individual features. DIN VDE 0100-300 [8] basically defines three types of power supply systems and the definitions:

- The TN- system
- The TT- system
- The IT-  system.

The lecture course should present the characteristic features of these systems according to the active electric conductor and corresponding type-related earth connection. The explanation of the sequence of letters identifying a network should be supported by pictures of electrical power networks (refer to DIN VDE 0100-300).
First letter, relation of the power supply system to earth:
    - T = direct connection of a point to earth
    - I = either all live parts are isolated from earth or a
        point is connected to earth via a large impedance.

Second letter, relation of the electrical installation to earth:
    - T = direct electrical connection of the body to earth
    - N = direct electrical connection of the body to an earthed point
        of the power supply system (neutral point).

## IEC Protection Classes

Electrical appliances are arranged into different electrical protection classes. The classification of the appliances is based on the safety precaution loosely titled "Protection from electric shock".

The European Standard EN 61140 [9] describes the various safety precaution measures and the four protection classes for electrically driven appliances. The IEC protection classes are:

- Class 0

  Apart from a basic insulation, there is no specific protection from an electric shock. It is not designed for the connection of an earth conductor.

- Class 1

  All the electrically conductive parts are connected to an earth conductor which is at earth potential.

- Class 2

  Class 2 appliances have a reinforced or double insulation between the mains power circuit and the output voltage. There is no connection to an earth conductor.

- Class 3

  Appliances in this class operate using SELV (safety extra low voltage), i.e. 50 V AC voltage or 120 V DC voltage.

An objective of the "Safety in Medical Engineering" lecture course must be to enable students to assign electrically powered devices and equipment to their corresponding protection classes and, thus, to determine and define any additional precautionary features which may be necessary. The graphic symbols associated with the respective protection classes are provided in IEC 60417 [10].

*Classification of Locations Used for Medical Purposes*

The classification of medical locations or rooms used for medical purposes is defined in DIN VDE 0100-710 [6]. The classification of the locations is based on the (electrical) safety of human beings, in particular of patients. The rooms or locations are classified into Groups 0 to 2 whereby Group 0 represents the highest level of (electrical) safety to patients.

Prospective medical engineers must consider the classification of the location groups from a professional point-of-view because the group classification assigned defines the setup and realisation of the electrical safety precautions and installations. One aim should be to sensitise students to the specific problems of electrical safety in medical buildings and locations.

Examples of locations should be used to explain and simplify the classifications to students. Examples of groups:

- Group 0
  Waiting rooms, patient reception, consulting rooms, minor treatment rooms
- Group 1
  Recovery rooms, endoscopy rooms, ECG rooms, delivery rooms, radiological diagnosis
- Group 2
  Operation theatres, intensive care units, cardiac catheter rooms.


*Currents and Leakage Currents in Medical Engineering*

Experience has shown that students find it difficult to distinguish and classify the wide range of different currents, particularly in respect of medical engineering. However, a basic requirement should be to address the following currents and types of current within the scope of a course of lectures on "Safety in Medical Engineering" and clarify them with the aid of practical examples.

The most important currents and an explanatory description for medical engineering purposes in accordance with DIN EN 62353 [4]:

- Earth Leakage Current
  The current which flows from the power adapter of one or more devices through the insulation to the PE conductor
- Device Leakage Current
  The current which flows from the power adapter through the PE conductor or exposed conductive parts of the housing and applied parts to earth
- Patient Auxiliary Current
  The current which flows from the patient connections through the patient to earth.

In addition, there are two important types of resistance and resistance values which must be addressed with regard to safety in medical engineering:

- Protective Earth Conductor Resistance
  The resistance between the conductive parts of a device which are connected to a PE conductor and earth contact in the power plug. It should tend towards 0 Ohm.
- Insulation Resistance
  The resistance of an insulation which is determined or measured with a defined AC voltage. It should tend towards infinite Ohm.

The previously mentioned currents and resistances should form the basis for theory regarding fault currents and currents in medical devices within the lecture course. They do not represent all the currents which occur in medical engineering. However, these basic currents, or fault currents, can be used to derive all the other significant currents. Experience has shown that students of medical engineering can come to terms with other types of currents quite easily after having being confronted with and understanding the basic currents.

*Electrical Safety Measurements*

The electrical safety measurements here relate to measuring the above mentioned currents and resistances using special measuring instruments developed for medical engineering purposes. After the currents and resistances have been theoretically covered during the lecture course, they should be measured in practice, i.e. on medical devices and equipment. Consolidation of measuring theory by laboratory experiments should have a high priority.

The objective of these practical experiments must be to familiarise students with handling and using the various measuring instruments and eliminate any reservations about measuring equipment and implementing it.

It is extremely useful to reproduce the measurement arrangements documented in DIN EN 60601-1-2 DIN EN 62353 in laboratory experiments.

*Safety-Related Controls*

After the students have received the theoretical foundations regarding the specifics of safety in medical engineering, they should be capable of completing a safety-related control in accordance with §6 of the Medical Devices Operator Ordinance (MPBetreibV). They should be told that technical safety controls (STK) contained are arranged as follows with similar test points:

- Name of the medical device
  Type, purpose, serial number, device manufacturer
- Data regarding the device operator, device owner
  Name, address
- Visual inspection of the medical device
  During the inspection, the housing, connected parts and all operating elements must be checked for visible signs of damage
- Function test
  During the function test, all the relevant output parameters related to this type of medical device are tested

- Electrical safety test

  During the electrical safety test, the parameters mentioned on Page 12 are measured and their actual values compared with the set values prescribed in the applicable standards.

Following the safety-related controls (STK), the student, the prospective medical engineer, should be capable of submitting an assessment or evaluation regarding the safety of the medical device tested. The test result is acknowledged by the tester applying the date of the test and his signature.

*Conclusion*

This concept was drawn up following the author's many years of experience in the field of lecturing, specifically in the field of medical devices and equipment.

The concept itself aims to represent a basic framework for training students of medical engineering. It does not claim to be a complete training concept. The main aim of this concept is to illustrate the most important basic aspects of the training course. This basic concept can be used as a framework within which to develop individual lectures and expand on them in related training and study programmes

Experience shows that the presentation of theoretical knowledge in lectures, particularly regarding the aspect of safety in medical engineering, is much easier to comprehend when accompanied by experiments and practical experience using medical devices in laboratory conditions.

Planning a course of lectures on the subject of "Safety in Medical Engineering" is not recommended if it is not possible to accompany the theory with practical experiments in laboratories.

This concept for a lecture course should be regarded as a "common thread" linking individual lectures within the scope of medical engineering studies.

*List of References (German literature)*

[1]    *Gesetz über Medizinprodukte (Medizinproduktegesetz-MPG)*
Medizinproduktegesetz in der Fassung der Bekanntmachung
vom 7. August 2002 (BGBl. I S. 3146), zuletzt geändert am
24.07.2010 (BGBl. I S.983)

[2]    *Verordnung über das Errichten, Betreiben und Anwenden von*
*Medizinprodukten* (Medizinprodukte-Betreiberverordnung-MPBetreibV)
in der Fassung der Bekanntmachung vom 21. August 2002 (BGBl. I s:
3396), zuletzt geändert am 29.07.2009 (BGBl. I S. 2326)

[3]    *DIN EN 60601-1-2; VDE 0750-1-2:2010-05*
Medizinische elektrische Geräte - Teil 1-2: Allgemeine Festlegungen
für die Sicherheit einschließlich der wesentlichen Leistungsmerkmale -
Ergänzungsnorm: Elektromagnetische Verträglichkeit - Anforderungen
und Prüfungen (IEC 60601-1-2:2007, modifiziert); DIN Deutsches
Institut für Normung e.V. und VDE Verband der Elektrotechnik
Elektronik Informationstechnik e.V. Berlin

[4]    *DIN EN 62353; VDE 0751-1:2008-08*
Medizinische elektrische Geräte - Wiederholungsprüfung und Prüfung
nach der Instandsetzung von medizinischen elektrischen Geräten (IE
62A/504/CDV:2005); DIN Deutsches Institut für Normung e.V. und
VDE Verband der Elektrotechnik Elektronik Informationstechnik e.V.
Berlin

[5]    *DIN VDE 0701-0702; VDE 0701-0702:2008-06*
Prüfung nach Instandsetzung, Änderung elektrischer Geräte -
Wiederholungsprüfung elektrischer Geräte - Allgemeine
Anforderungen für die elektrische Sicherheit. DIN Deutsches Institut
für Normung e.V. und VDE Verband der Elektrotechnik Elektronik
Informationstechnik e.V. Berlin

[6]     *DIN VDE 0100-710;2004-06*
        Errichten von Niederspannungsanlagen - Anforderungen für
        Betriebsstätten, Räume und Anlagen besonderer Art - Teil 710:
        Medizinisch genutzte Bereiche (IEC 60364-7-710:2002, modifiziert).
        DIN Deutsches Institut für Normung e.v. und VDE Verband der
        Elektrotechnik Elektronik Informationstechnik e.v. Berlin

[7]     *BGV A3 Berufsgenossenschaftliche Vorschriften*
        Unfallverhütungsvorschrift / Durchführungsanweisungen
        Elektrische Anlagen und Betriebsmittel
        Ausgabe: 2005-01; HVBG Hauptverband der gewerblichen
        Berufsgenossenschaften in Kooperation mit dem Carl Heymanns
        Verlag 2005

[8]     *DIN VDE 0100-300, 1996-01*
        Errichten von Starkstromanlagen mit Nennspannungen bis 1000V,
        Teil 3: Bestimmungen allgemeiner Merkmale (IEC 364-3:1993,
        modifiziert). DIN Deutsches Institut für Normung e.v. und VDE
        Verband der Elektrotechnik Elektronik Informationstechnik e.v. Berlin

[9]     *DIN EN 61140 (VDE 0140-1); 2007-03*
        Schutz gegen elektrischen Schlag- Gemeinsame Anforderungen für
        Anlagen und Betriebsmittel (IEC 61140:2001+A1:2004, modifiziert).
        DIN Deutsches Institut für Normung e.v. und VDE Verband der
        Elektrotechnik Elektronik Informationstechnik e.v. Berlin

[10]    *IEC 60417:1973 + IEC 417A:1974 bis IEC 417M:1994*
        Grafische Symbole für Einrichtungen (Bildzeichen).
        International Electrotechnical Commission, Genf.
        Beuth Verlag GmbH